W9-ACW-612

ANSWER ME THIS, WORLD BOOK

Do penguins have emotions?

World Book
answers YOUR questions
- about -
the oceans and what's in them

www.worldbook.com

The questions in this book came from curious kids just like you who want to make sense of the world. They wrote to us at World Book with nothing but a question and a dream that the question they've agonized over would finally be answered.

Some questions made us laugh. Others made us cry. And a few made us question absolutely everything we've ever known. No matter the responses they induced, all the questions were good questions.

There isn't a strict rule for what makes a question good. But asking any question means that you want to learn and to understand. And both of those things are very good.

Adults are always asking, "What did you learn at school today?" Instead, we think they should be asking, **"Did you ask a good question today?"**

Why is the

ocean blue?

Just kidding, natural elements don't fight over colors...that we know of. Plus, the ocean can appear green sometimes, too.

But most people say the ocean is blue. It appears blue due to how ocean water absorbs sunlight. Light from the sun is made up of all colors of the rainbow. Water absorbs red, orange, yellow, and green light. But not blue. The blue parts are scattered and reflected widely. So that's why the ocean looks blue.

What's the smallest shark?

And how many sharks are there?

Itty-bitty.

Or should we say itty-**bitey.** The smallest shark is the dwarf lanternshark, which can be about six inches (16 centimeters) long and weigh one ounce (28 grams). That's about the height and weight of a slice of bread. There are 400 species of sharks, many varying in size.

What are sea cucumbers?

They're an important ingredient in an underwater Greek salad!

While that would be great news for salads, it's not true. Sea cucumbers live in oceans and belong to a group of spiny-skinned animals called *echinoderms*. Hundreds of kinds of sea cucumbers live throughout the oceans at all depths. Some tropical sea cucumbers grow 2 to 3 feet (60 to 90 centimeters) long. But some people do eat sea cucumbers on land. In Asia, sea cucumbers are dried and sold as food called *trepang*.

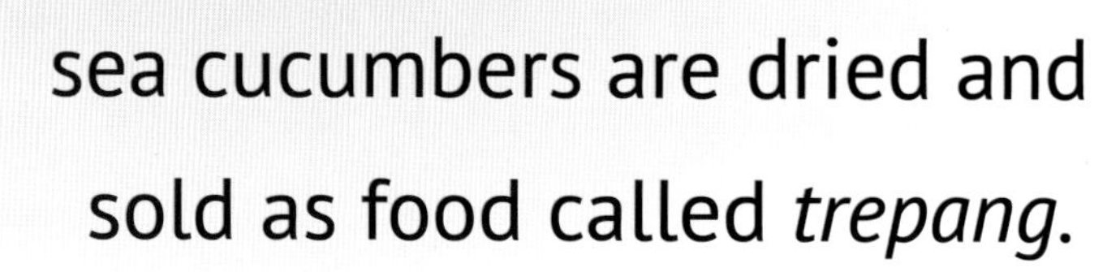

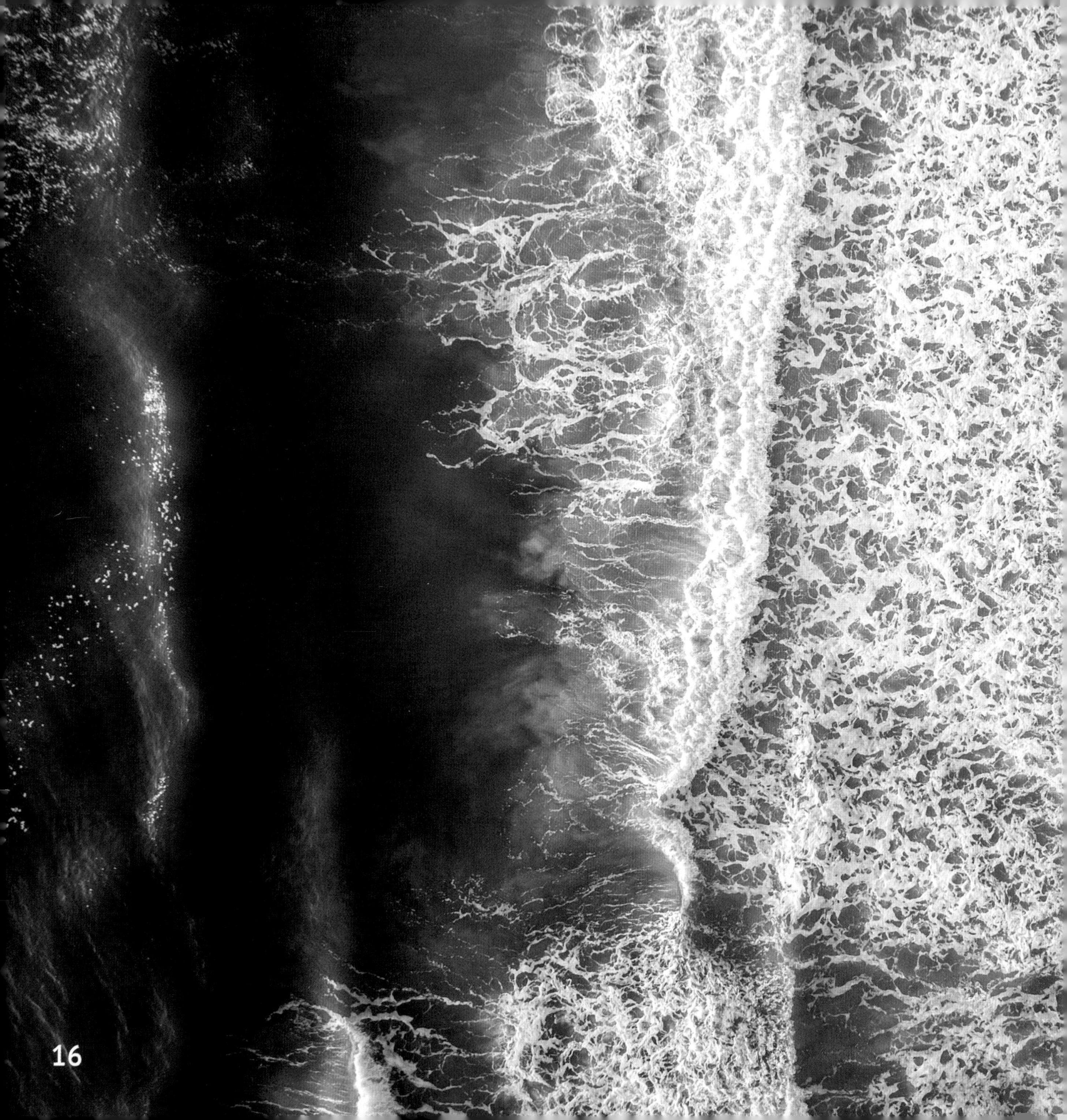

Why is the ocean salty?

Because the salt needs to balance out the pepper.

Didn't you know that's what sand is? No, we're joking, that isn't true. But this is: the ocean is salty because it rocks. The saltiness of the ocean comes from rocks breaking down on land. As rocks break down, rivers carry the salty bits of rock to the oceans where they dissolve. The saltiness is concentrated as fresh water evaporates from the ocean's surface.

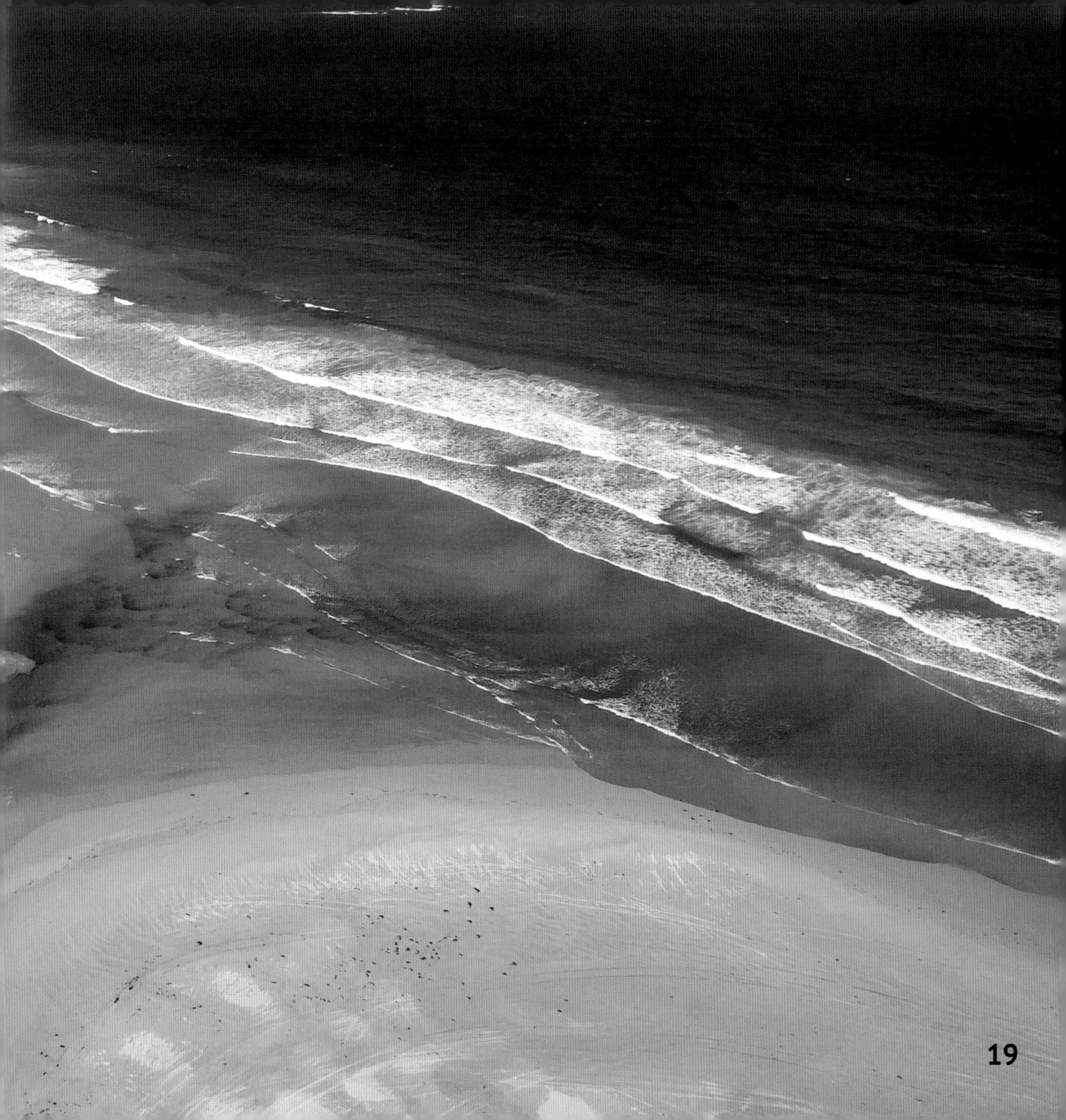

What is the heaviest animal on Earth?

126.0 kg

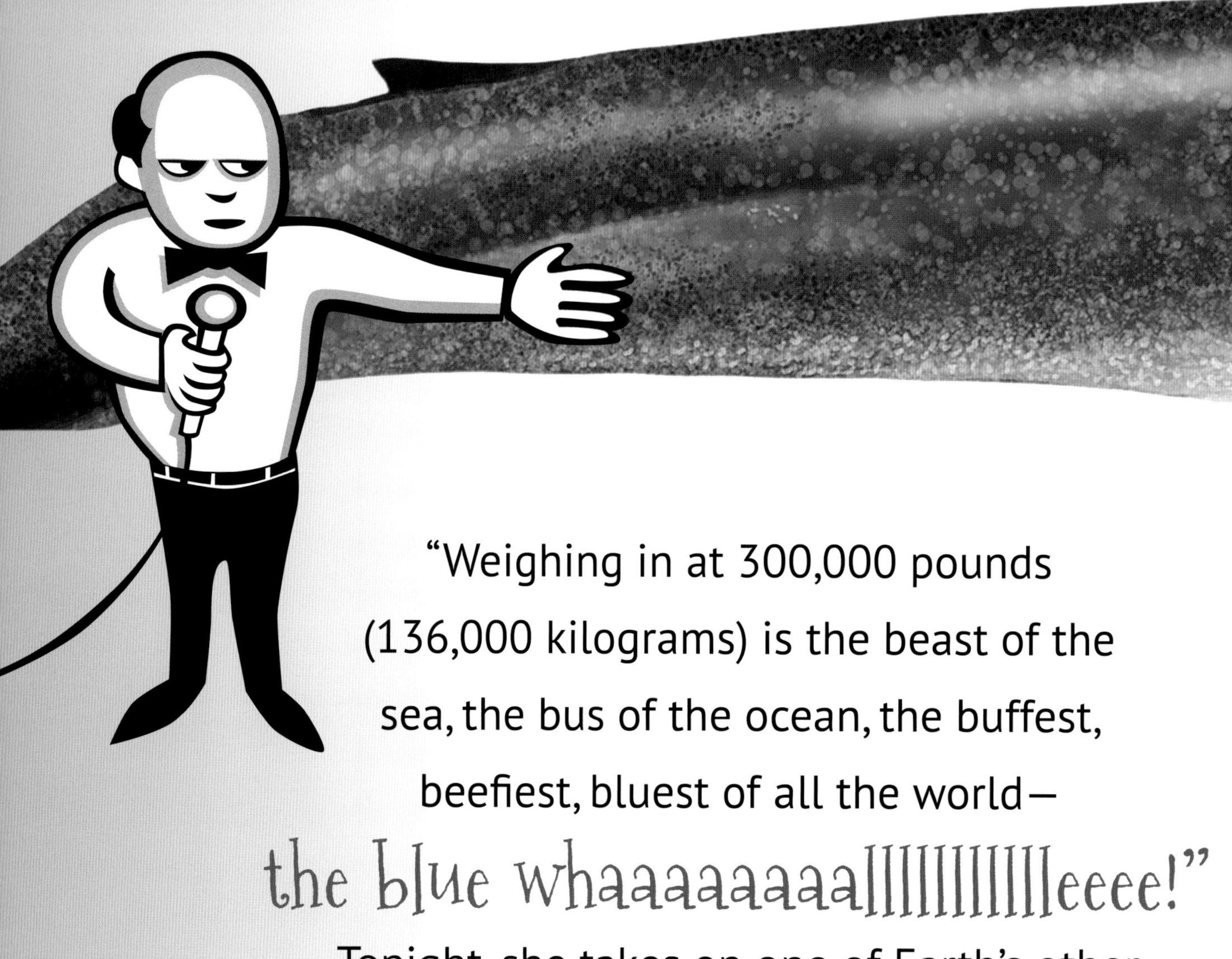

"Weighing in at 300,000 pounds (136,000 kilograms) is the beast of the sea, the bus of the ocean, the buffest, beefiest, bluest of all the world—

the blue whaaaaaaaaallllllllllleeee!"

Tonight, she takes on one of Earth's other behemoths: the African elephant, the largest living animal that walks on land!

But the elephant is in for a rough ride—it weighs only as much as the blue whale's tongue!" (Ding, ding.) "Let's get ready to

rrruuummmbbblllleee!"

Why do sharks have rough skin?

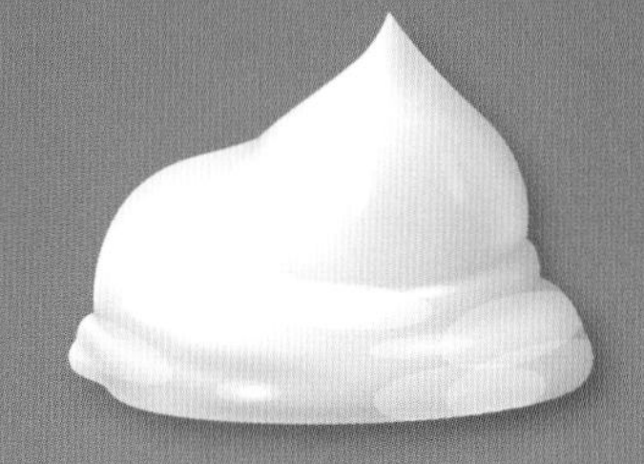

They never moisturize.

But they should. The small, tooth-like scales all over them make them have rough skin. The sharp scales are called *placoid* scales. Rough skin helps protect sharks from other fish bumping into them—or trying to take a bite. Dried sharkskin is so rough that it was once used as sandpaper.

Why are seals and sea lions black?

It may seem strange,

but not all seals and sea lions are black. There are more than 30 kinds of seals and sea lions. They range in color from white to gray to brown to black. Some are even spotted or have a pattern on them that is two or more colors.

How polluted are oceans?

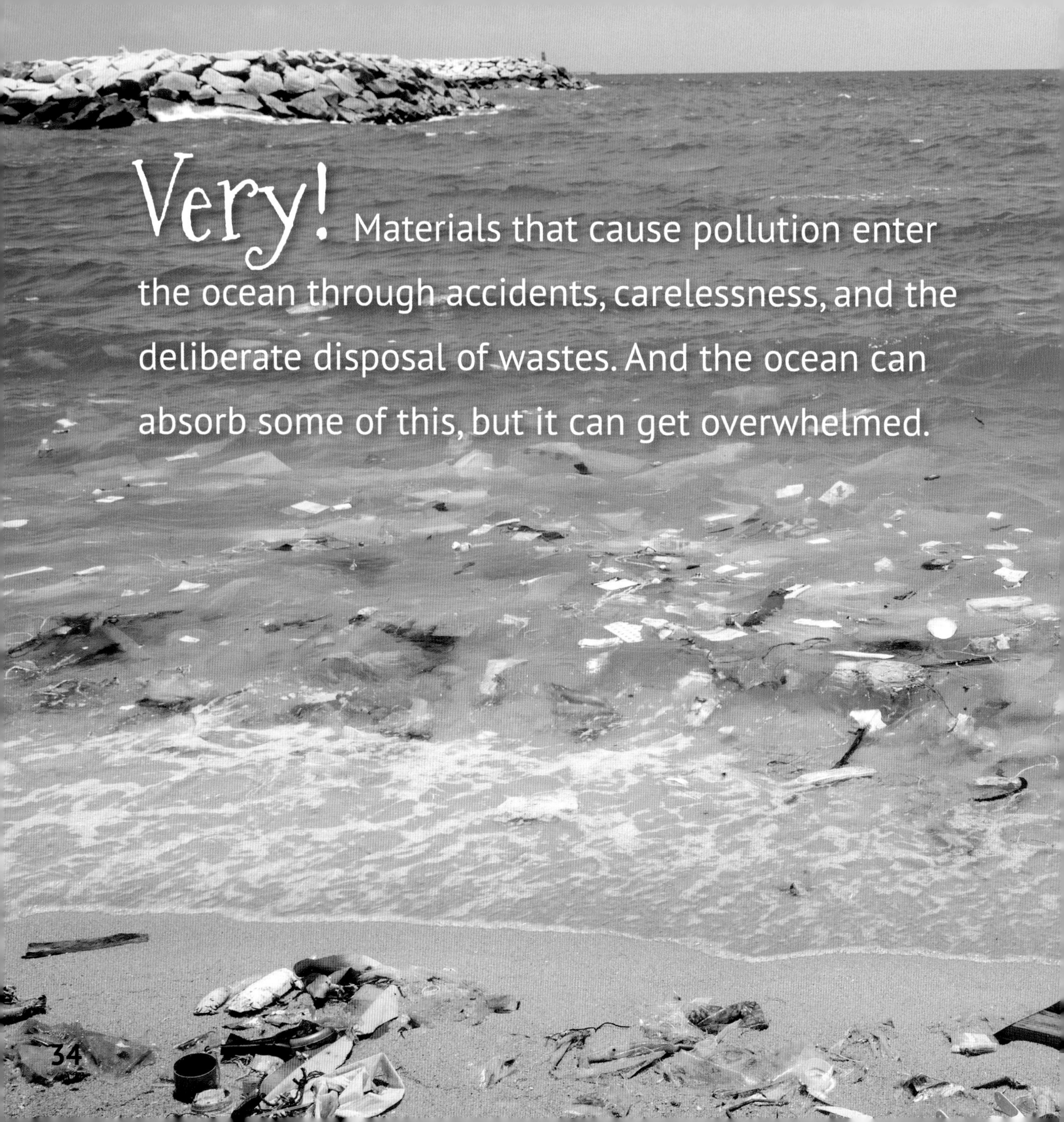

Very! Materials that cause pollution enter the ocean through accidents, carelessness, and the deliberate disposal of wastes. And the ocean can absorb some of this, but it can get overwhelmed.

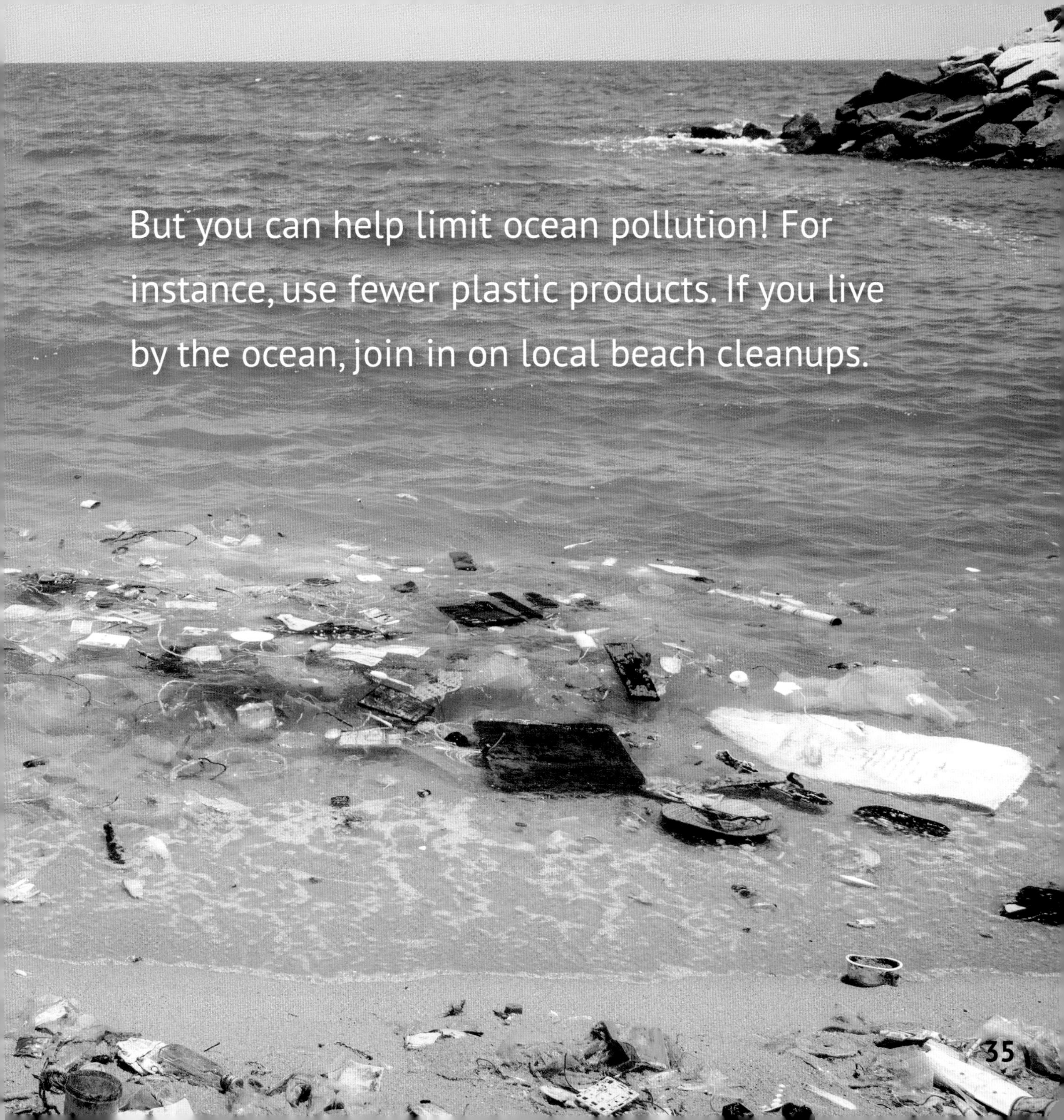

But you can help limit ocean pollution! For instance, use fewer plastic products. If you live by the ocean, join in on local beach cleanups.

How big was megalodon?

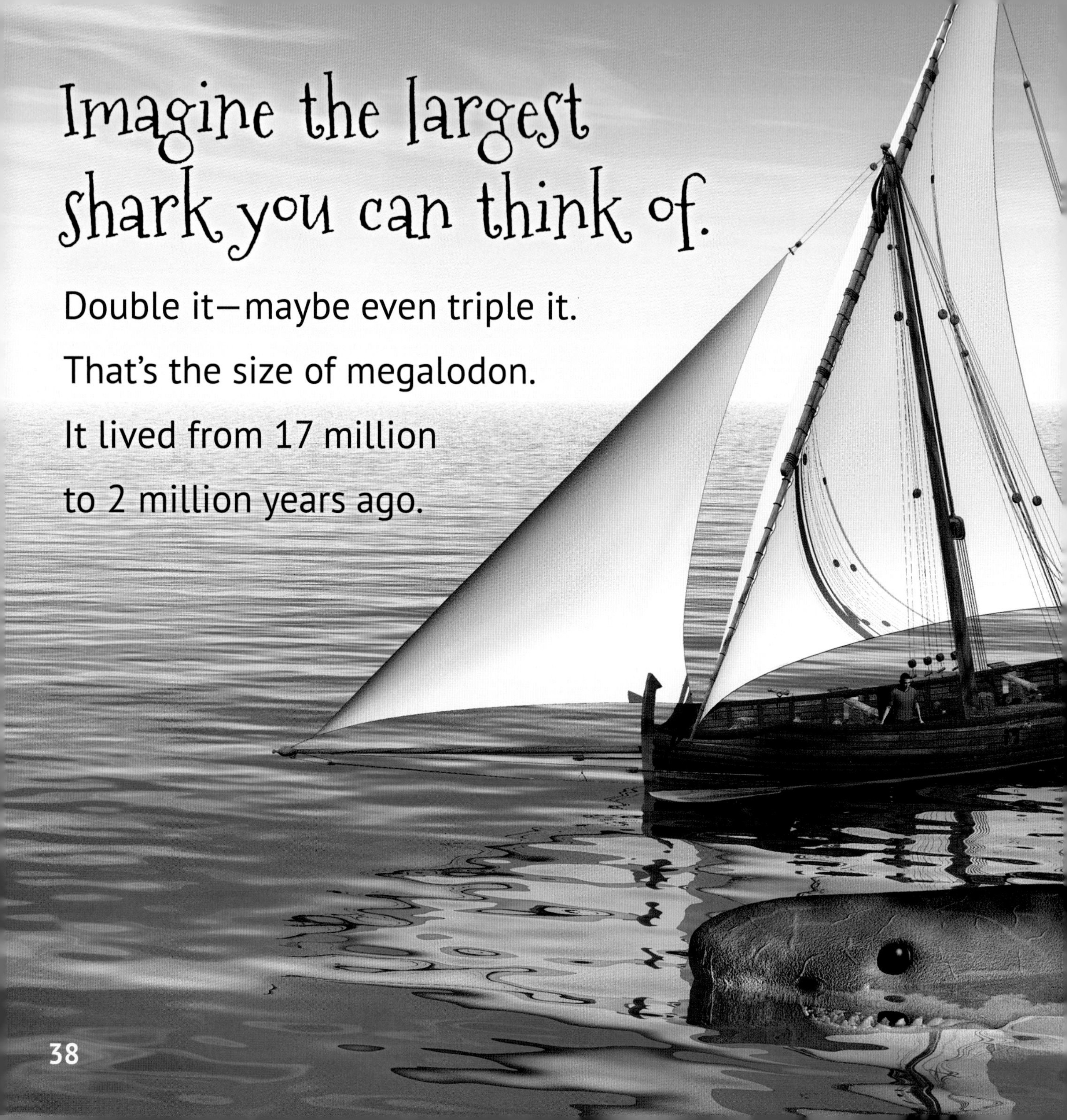

Imagine the largest shark you can think of.

Double it—maybe even triple it.

That's the size of megalodon.

It lived from 17 million

to 2 million years ago.

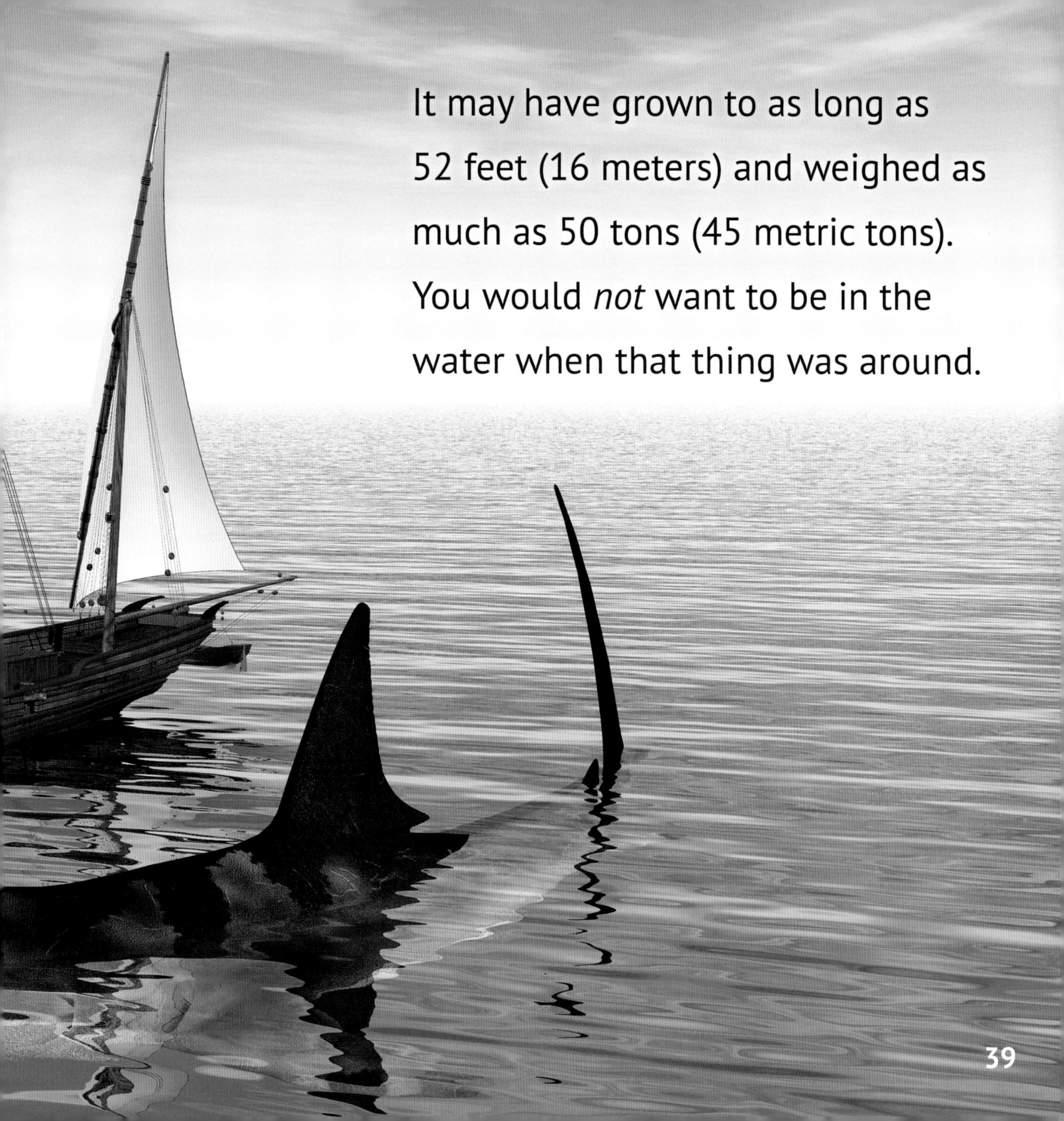

It may have grown to as long as 52 feet (16 meters) and weighed as much as 50 tons (45 metric tons). You would *not* want to be in the water when that thing was around.

How are sea turtles born?

It's a Secret.

A mama sea turtle lays eggs on land. Heat from the sun keeps the eggs warm until they are ready to hatch. This is where the secret part comes in: once hatched, baby turtles must dig their way to the surface and make it to the ocean. On the way, predators try to snatch them up, so they have to be very fast. Sometimes to protect them, humans shield the hatchlings. This makes their journey safe and like a secret from predators.

How salty does the ocean have to be for an egg to float?

On a scale of pretzel to anchovy, the ocean has to be at about a potato chip.

If that unscientific scale didn't help you, here's a better answer: an egg will float in any ocean because all ocean water is salty. Oceans are at least 4 to 6% dissolved salts by weight. So it doesn't need to be any saltier than it already is. If you yourself want to float, you can go to the Dead Sea in the Middle East. It's about 25 to 30% dissolved salts. That's most definitely anchovy level.

Why do fish have scales?

Imagine a knight galloping through a field.

Arrows lightly kiss his armor as he falls to the ground. His armor protects him from danger. It's the same thing with a fish—only underwater. And the arrows are other fish. Fish scales protect the skin from predators and injuries. They also resist water, so a fish can zip and zoom through the water—almost like it's riding a horse.

How deep is the deepest

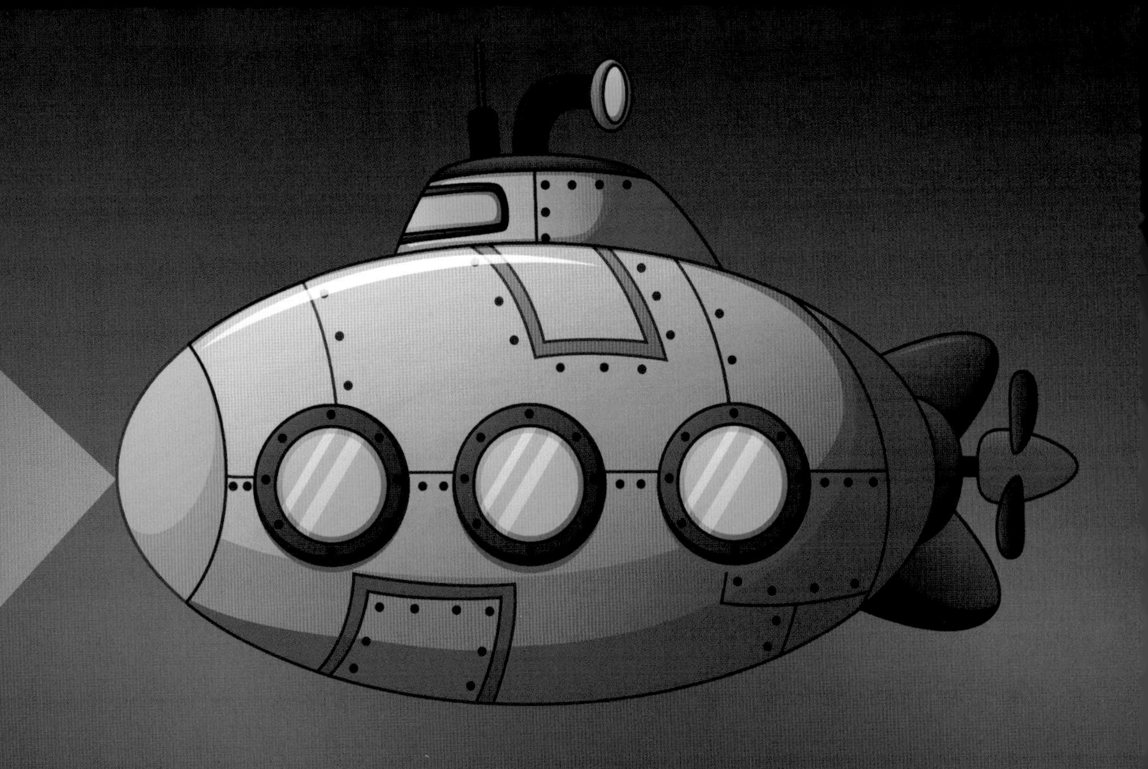

part of the ocean?

You know how Mount Everest is super tall?

At 29,035 feet (8,850 meters), it's the world's highest mountain. Well, if you put Mount Everest in the deepest part of the ocean, more than 1.3 miles (2.1 kilometers) of water would cover the mountaintop. That was a very long way of saying that the deepest part of the ocean is the Mariana Trench in the Pacific Ocean. It lies 35,840 feet (10,924 meters) below sea level. We had to make you dive for the answer.

Mount Everest
Mariana Trench

What is the Bermuda Triangle?

We had a better answer to this question, but it disappeared in the Bermuda Triangle.

Because that's what happens in the Bermuda Triangle—things seem to disappear. It is an area of the ocean off the southeastern coast of Florida. Many ships and airplanes have disappeared while traveling in that area. Some people think supernatural forces are at work...but there is also a perfectly reasonable explanation to all of this. Scientists think that violent, unexpected storms or strong air currents may have destroyed the wreckage from ships and planes.

Are killer whales

dangerous?

Of course killer whales are dangerous!

It's in the name: killer! But, wait, we're letting our extreme fear of sharks get in the way. Killer whales, also called orcas, feed on fish, squids, and marine mammals like dolphins, porpoises, seals, and other whales. *But* orcas don't normally attack people.

How does such a pretty pearl come from such an ugly thing?

It's very irritating!

When an irritant, such as a parasite or bit of rough sand, enters the body of an oyster, a special organ gets to work. This fleshy organ is called the nacreous layer. It produces thin sheets of nacre—shiny stuff. The oyster starts layering and layering the invader with thin sheets of smooth nacre. Kind of like layering and layering pearl bracelets. The shiny nacre coats all around until the foreign object is enclosed. That enclosed foreign object is your pearl. Now you only need about 80 more to add another bracelet to your arm.

Is it possible to raise and rebuild

the *Titanic?*

You know how people often say, "Anything is possible"?

In this instance, that isn't true. The *Titanic* has been under water for more than 100 years. It's broken into many, many pieces. The steel plate pieces would fall apart once they hit the air. That means they would finally get to see what this world is like a century after they sank, only to quickly fall apart. Rumor has it a *replica* (copy) of the *Titanic* is going to set sail one day!

Why can I hear the ocean in a seashell?

You can't!

Hearing the ocean when you hold a seashell to your ear is a popular myth—like a lot of questions we answer in this series! When you hold a seashell to your ear, you hear ambient noise. That means the noise in the area *around* where you're trying to hear seagulls and the grunts from beach volleyball players. But, it's still fun to pretend like you're at the beach. Treat it like a staycation!

Do penguins have

emotions?

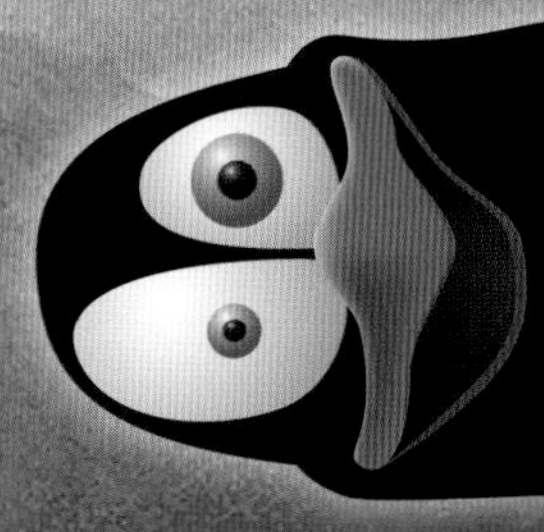

(sniffle, sniffle)
"Ye-Yes! We d-d-d-do have emotions!

Yesterday, my peng—I mean ex-penguin!—
—broke up with me. Get out of here! Fly away!
Oh, that's right, you can't! Because you're a
penguin! (sniffle)...We have emotions just
like humans, we just experience and express
feelings differently than you would. (sniffle) You
can probably tell from my weepy tears, hanging
head, and mournful sounds that I am sad.
I th-th-think (sniffle) a box of chocolates would
make me feel better...or a smelly, fresh fish."

Is there really a lost city in the ocean?

Maybe.

That lost city you're asking about is called Atlantis. It was described in legends. People believe that it sank to the bottom of the ocean. But nobody has found it …yet. It probably did not really exist. One person who made the idea of Atlantis popular is named Plato. And, no, he didn't invent the fun, colorful goop you play with. He was a philosopher in ancient Greece who came up with a lot of smart ideas. And here's a bad idea: build a boat out of that fun, colorful goop and try to find Atlantis.

What would you bring to a desert island?

A survival expert!

She would know what to do. But that's missing the point. To survive on a deserted island, it would be important to be able to build a shelter, get safe drinking water, and gather and cook food. Fire would keep you warm, help you purify water, keep predators away, and cook your food. So you'd want matches or a piece of flint to help start a fire. A knife or machete would help cut down materials for a shelter and hunt for food. And a cell phone to call for help would be nice...

N
NE
E
SE
S
SW
W
NW

What's the biggest shark?

Can you picture two elephants stacked on top of each other traveling through the ocean?

If yes, you have a very active imagination. And that's the size of the largest shark: the whale shark. They may weigh over 15 tons (14 metric tons). The whale shark is not only the biggest shark, but it is the largest living fish. That's quite an accomplishment. Even though they are huge, whale sharks are harmless to people. Phew!

Where and when do polar bears hibernate?

Do
not
disturb

Polar bears live in coastal areas of Alaska, Canada, Greenland, Norway, and Russia.

While a lot of bears like to catch some z's in the winter, most polar bears stay active. But pregnant polar bears usually stay in a den from October to March or April. They dig dens in deep snowbanks on the side of a hill or valley. (yawn) Writing a book is exhausting—we better catch some z's, too.

World Book, Inc.
180 North LaSalle Street
Suite 900
Chicago, Illinois 60601
USA

For information about other "Answer Me This, World Book" titles, as well as other World Book print and digital publications, please go to www.worldbook.com.

For information about other World Book publications, call 1-800-WORLDBK (967-5325).

For information about sales to schools and libraries, call 1-800-975-3250 (United States) or 1-800-837-5365 (Canada).

Library of Congress Cataloging-in-Publication Data for this volume has been applied for.

Answer Me This, World Book
ISBN: 978-0-7166-3821-6 (set, hc.)

Do penguins have emotions?
World Book answers your questions about the oceans and what's in them
ISBN: 978-0-7166-3823-0 (hc.)

Also available as:
ISBN: 978-0-7166-3833-9 (e-book)

Printed in China by RR Donnelley,
Guangdong Province
1st printing July 2019

Acknowledgments

Cover © Alexey Suloev, Shutterstock; © Elena Akkurt, Shutterstock
3-95 © Shutterstock

Editorial

Writers
Madeline King
Grace Guibert

Manager, New Content Development
Jeff De La Rosa

Manager, New Product Development
Nick Kilzer

Proofreader
Nathalie Strassheim

Manager, Contracts and Compliance (Rights and Permissions)
Loranne K. Shields

Manager, Indexing Services
David Pofelski

Digital

Director, Digital Product Development
Erika Meller

Digital Product Manager
Jonathan Wills

Graphics and Design

Senior Visual Communications Designer
Melanie Bender

Media Editor
Rosalia Bledsoe

Manufacturing/Production

Manufacturing Manager
Anne Fritzinger

Production Specialist
Curley Hunter